Diodes & BJTs

By

Gabriel Alfonso Rincón-Mora

School of Electrical and Computer Engineering
Georgia Institute of Technology

Rincon-Mora.gatech.edu

Contents

List of Figures

List of Abbreviations

BJT $\equiv$ Bipolar-Junction Transistor

FET $\equiv$ Field-Effect Transistor

MOS $\equiv$ Metal–Oxide–Semiconductor

A_J $\equiv$ Junction Area

α_T $\equiv$ Base-Transport Factor

α_0 $\equiv$ Baseline Transport Factor

β_0 $\equiv$ Baseline Base–Collector Current Gain

C_{DEP} $\equiv$ Depletion Capacitance

C_{DIF} $\equiv$ Diffusion Capacitance

C_J $\equiv$ Junction Capacitance

C_{J0} $\equiv$ Zero-Bias Junction Capacitance

D_N $\equiv$ Electron Diffusion Coefficient

D_H $\equiv$ Hole Diffusion Coefficient

e^- $\equiv$ Electron

E_B $\equiv$ Energy Barrier

E_{BG} $\equiv$ Band-Gap Energy

E_C $\equiv$ Conduction-Edge Energy

E_E $\equiv$ Electron Energy

E_F $\equiv$ Fermi Level

E_V $\equiv$ Valence-Edge Energy

g_m $\equiv$ Small-Signal Transconductance

γ_E $\equiv$ Emitter Injection Efficiency

h^+ $\equiv$ Hole (Missing Electron)

$i_B \equiv$ Base Current

$i_C \equiv$ Collector Current

$i_D \equiv$ Diode Current

$i_E \equiv$ Emitter Current

$i_F \equiv$ Forward Diode Current

$i_R \equiv$ Reverse Diode Current

$i_{RC} \equiv$ Recombination Current

$I_S \equiv$ Reverse-Saturation Current

$K_B \equiv$ Boltzmann's Constant

$L_N \equiv$ Electron's Average Diffusion Length

$L_P \equiv$ Hole's Average Diffusion Length

$\mu_N \equiv$ Electron Mobility

$\mu_P \equiv$ Hole Mobility

$n_E \equiv$ Electron Density

$n_H \equiv$ Hole Density

$n_I \equiv$ Intrinsic Carrier Concentration/Non-Ideality Factor

$N_A \equiv$ Acceptor Doping Concentration

$N_B \equiv$ Base Doping Concentration

$N_C \equiv$ Collector Doping Concentration

$N_D \equiv$ Donor Doping Concentration

$N_E \equiv$ Emitter Doping Concentration

$q_E \equiv$ Electronic Charge

$q_{FR} \equiv$ Forward-Recovery Charge

$q_{RR} \equiv$ Reverse-Recovery Charge

$t_{FR} \equiv$ Forward-Recovery Time

$t_R \equiv$ Recovery Time

$t_{RR} \equiv$ Reverse-Recovery Time

$T_J \equiv$ Junction Temperature

$\tau_F \equiv$ Forward Transit Time

$\tau_H \equiv$ Hole's Average Carrier Lifetime

$\tau_N \equiv$ Electron's Average Carrier Lifetime

$v_B \equiv$ Base Voltage

$v_{BC} \equiv$ Base–Collector Voltage

$v_{BE} \equiv$ Base–Emitter Voltage

$v_C \equiv$ Collector Voltage

$v_{CE} \equiv$ Collector–Emitter Voltage

$v_{CE(MIN)} \equiv$ Minimum Collector–Emitter Voltage

$v_D \equiv$ Diode Voltage

$v_E \equiv$ Emitter Voltage

$v_R \equiv$ Reverse (Diode) Voltage

$V_{BD} \equiv$ Breakdown Voltage

$V_{BI} \equiv$ Built-In Potential Voltage

$V_t \equiv$ Thermal Voltage

$w_B \equiv$ Effective Base Width

$w_{B0} \equiv$ Zero-Bias Base Width

$W_B \equiv$ Metallurgical Base Width

Diodes & BJTs

Power supplies use switches to draw, steer, and deliver charge from input sources into rechargeable batteries and microelectronic loads. Semiconductor companies use *diodes, bipolar-junction transistors* (BJTs), and *complementary metal–oxide–semiconductor* (CMOS) *field-effect transistors* (FETs) for this purpose. Of these, FETs are oftentimes preferable because they drop lower voltages than diodes and require less current to switch than BJTs, so they consume less power. Still, diodes do not require a synchronizing signal like FETs and BJTs, and BJTs cost less money. Plus, MOSFETs incorporate diodes and BJTs that can at times activate and steer some or all of the current. So understanding how diodes and BJTs conduct current is essential.

1. Solids

1.1. Energy-Band Diagram

Electrons in populated orbits of a material are bound to their home sites around the nucleus in Fig. 1. Electrons in the outermost orbit are responsible for bonding with other atoms. These are the *valence electrons* that populate the *valence band* and form covalent bonds. Electrons that break free are available for conduction. These free *charge carriers* are in the *conduction band*.

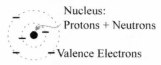

Nucleus:
Protons + Neutrons

Valence Electrons

Fig. 1. Atom.

Although available for conduction, *electron affinity* keeps these electrons in their orbits. Combined potential and kinetic energy E_E is lower for tightly bound electrons. So *electron energy* E_E in the conduction band in Fig. 2 is greater than in the valence band. *Conduction-edge energy* E_C is therefore the minimum energy that liberated electrons carry. *Valence-edge energy* E_V is similarly the maximum energy that bound valence electrons hold.

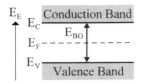

Fig. 2. Energy-band diagram.

Electrons orbit around the nucleus at discrete energy levels. So no electrons reside between the valence and conduction bands. The energy span between these two bands is the *band gap*. This *band-gap energy* E_{BG} is the energy needed to liberate and promote valence electrons into the conduction band. E_{BG} is therefore the difference between E_C and E_V.

1.2. Conduction

Electrons that rise into the conduction band leave voids in the valence band. Once liberated, these free electrons drift easily. Neighbor valence electrons also shift easily into valence *holes*. And as these valence electrons shift positions, holes drift in the opposite direction. Holes in the valence band therefore carry charge like electrons in the conduction band.

Since liberated electrons create holes, the probability of finding electrons in the conduction band of homogeneous material is equal to the probability of finding holes in the valence band. Because free electrons reside in the conduction band and holes in the valence band, the most probable energy level for a charge carrier (when neglecting that the band gap excludes electrons and holes) is halfway between the bands. And, the probability that this charge carrier is an electron or a hole is 50%.

This 50% probability is what the *Fermi level* E_F indicates and why E_F in homogeneous material is halfway between E_C and E_V. The probability of finding charge carriers above and below this level falls exponentially. E_F is effectively an indicator of charge-carrier density. And like water in a lake, charges do not flow when the concentration across a material is uniform. So E_F is uniform across a material when current is zero and sloped otherwise.

1.3. Classification

Conductors like metal and aqueous solutions of salts conduct charge easily because valence electrons are so weakly bound to their home sites that they are practically free and available for conduction. This is another way of saying that the valence band overlaps the conduction band. Valence electrons in *insulators* like rubber and plastic, on the other hand, require so much band-gap energy to liberate that they hardly conduct. The band gap in *semiconductors* like *silicon* Si and *germanium* Ge is moderate, so they conduct moderately well. In silicon, the band gap energy is 1.1 eV or 1.8×10^{-19} J at 27° C.

1.4. Semiconductors

Thermally excited electrons that break free from their valence positions avail electron–hole pairs that can carry charge. These *thermionic emissions* produce an equal number of holes and electrons. This is the *intrinsic carrier concentration* n_i of a semiconductor. This concentration is higher when the band-gap energy that binds valence electrons is lower and the temperature that energizes them is higher. n_i for silicon at 27° C is 1.45×10^{10} cm^{-3}.

Hole density n_H and *electron density* n_E in *intrinsic semiconductors* equal this n_i because these materials are homogeneous and pure. *Dopant atoms* with partially populated outer orbits are impurities that can alter these concentrations. So *doped semiconductors* are semiconductors with atomic impurities that produce uneven carrier concentrations.

A. N Type

Electrons in the outer orbit of *donor atoms* are so loosely bound in doped semiconductors that they are free in the conduction band. These dopant atoms effectively "donate" negatively charged electrons e^-'s. This is why engineers say material doped with donor atoms is negative or *N type*. In spite of this appellation, the material is electrically neutral because the intrinsic and dopant atoms that comprise it are neutral.

With more electrons in the conduction band, the probability that thermionic holes in the valence band recombine is higher. Electron and hole carrier concentrations are lower as a result. n_H, for example, reduces to the number of thermionic holes n_h that do not recombine. n_h is therefore lower than n_i by the amount that *donor doping concentration* N_D dictates:

$$n_H = n_h = \frac{n_i^2}{N_D} < n_i, \tag{1}$$

which is what the *mass-action law* states.

The situation for n_E is different because dopants add N_D electrons to the conduction band. Plus, N_D is normally orders of magnitude higher than n_i. So N_D is so much higher than the number of thermionic electrons that do not recombine n_e that n_E is nearly N_D:

$$n_E = N_D + n_e \approx N_D, \tag{2}$$

N_D therefore determines the extent to which the material is N type. n_h is so much lower than the resulting n_E that holes in N-type material are *minority carriers* and electrons are *majority carriers*.

Example 1: Find n_H at 27° C is silicon when doped with $10^{14}/cm^3$ donor atoms.

Solution:

$$n_H = n_h = \frac{n_i^2}{N_D} = \frac{(1.45 \times 10^{10})^2}{10^{14}} = 2.10 \times 10^6 cm^{-3}$$

When doped this way, the probability of finding free *electrons* e^-'s is higher than that of finding *holes* h^+'s. So E_F in N-type material is closer to the conduction band in Fig. 3 than to the valence band. Since the probability of finding charge carriers drops exponentially away from E_F and the band gap is free of carriers, n_E peaks at the edge of the conduction band and decreases exponentially above E_C. Although to a lower extent, n_H similarly peaks at the edge of the valence band (where electrons are more likely to break free) and decreases exponentially below E_V.

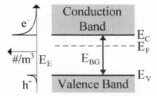

Fig. 3. Band diagram of N-type semiconductors.

B. P Type

Acceptor atoms produce the opposite effect. Electrons in the outermost orbit of acceptor atoms are so tightly bound in doped semiconductors that they are in the valence band. This outer orbit is incomplete, however, with electron vacancies or holes h^+'s. Since these impurities are more likely to "accept" than donate electrons, engineers say material doped with acceptor atoms is positive or *P type*. The material is nevertheless electrically neutral because the intrinsic and dopant atoms that comprise it are neutral.

With more holes in the valence band, the probability that thermionic electrons in the conduction band recombine is higher, so n_E and n_H are lower. n_E therefore reduces to the number of thermionic electrons n_e that do not recombine. n_e is lower than n_i by the amount that *acceptor doping concentration* N_A dictates:

$$n_E = n_e = \frac{n_i^2}{N_A} < n_i. \tag{3}$$

Since dopants add N_A holes to the valence band and N_A is normally orders of magnitude greater than n_i, N_A holes overwhelm the thermionic holes that do not recombine n_h. As a result, n_H is nearly N_A:

$$n_H = N_A + n_h \approx N_A. \tag{4}$$

N_A determines the extent to which the material is P type this way. n_e is so much lower than this n_H that free electrons are minority carriers in P-type material and holes are majority carriers.

When doped this way, the probability of finding holes is higher than that of electrons. So E_F in P-type material is closer to the valence band in Fig. 4 than to the conduction band. Since the probability of finding charge carriers drops exponentially away from E_F and the band gap is free of carriers, n_H peaks at the edge of the valence band and decreases exponentially below E_V. Similarly, but to a lower extent, n_E peaks at the edge of the conduction band and decreases exponentially above E_C.

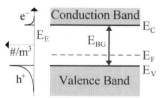

Fig. 4. Band diagram of P-type semiconductors.

C. Mobility

Carrier mobility is the ease with which carriers flow when exposed to an electric field. It increases with temperature because thermal energy excites electrons into more mobile states. This higher kinetic energy eases their movement, and with it, conduction. Bound valence electrons shift into holes in the valence band with less ease than free electrons drift in the conduction band. Valence and nucleic bonds are to blame for this. The effective nucleic mass of holes is therefore higher than that of free electrons. As a result, *hole mobility* μ_P is two to three times lower than *electron mobility* μ_N.

D. Notation

Superscripted plus and minus signs normally indicate relative concentration levels. So N_D^+, N_D, and N_D^-, respectively, produce heavily, moderately, and lightly doped N-type material that engineers denote with N^+, N, and N^-. N_D^+ is also usually orders of magnitude greater than N_D and N_D is similarly greater than N_D^-.

The same applies to P-type semiconductors. N_A^+, N_A, and N_A^- produce heavily, moderately, and lightly doped material P^+, P, and P^-. Unless otherwise specified, N_D^+ in N^+ is usually on the same order of magnitude as N_A^+ in P^+, and likewise for N_D in N and N_A in P and N_D^- in N^- and N_A^- in P^-. When doping concentration is so high that ohmic resistance is comparable to metal, the semiconductor is *degenerate*. In other words, degenerate semiconductors are good ohmic contacts.

2. PN-Junction Diodes

A *PN-junction diode* is a piece of semiconductor doped so acceptor impurity atoms outnumber donor impurity atoms on one side and *vice versa* on the other. The material transitions from one type to the other at the *metallurgical junction* X_J shown in Fig. 5. This is where doping concentrations effectively cancel. The doping difference $N_A - N_D$ transitions across this zero point abruptly in *step junctions* and more gradually in *linearly graded junctions*.

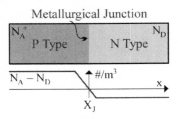

Fig. 5. P⁺N junction.

2.1. Zero Bias

A. Electrostatics

<u>Diffusion</u>: *Diffusion* is the force in nature that impels motion from dense regions to sparse spaces. In a PN junction, holes in the P side outnumber holes in the N side by orders of magnitude. Electron density in the N side is also much greater than electron density in the P side. Majority carriers therefore diffuse across the junction: holes to the N side and electrons to the P side.

<u>Depletion</u>: Diffusing electrons eventually populate holes in the P side when the system reaches *thermal equilibrium*. Diffusing holes similarly capture free electrons in the N side. This recombination process depletes the region near the junction of charge carriers. This carrier-free space in Fig. 6 is the *depletion region*.

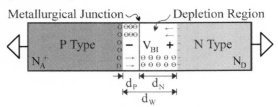

Fig. 6. Zero-bias P⁺N junction.

<u>Ionization</u>: Parent atoms lose and receive charge as carriers diffuse in and out of their orbits. Departing holes and incoming electrons charge the P side negatively and departing electrons and incoming holes charge the N side positively. So the P side begins to repel incoming electrons and the N side to repel incoming holes.

Charge carriers nevertheless continue to diffuse until the electric field is strong enough to repel further action, which happens when carrier density (and E_F) is uniform across the junction. The field that results across this *space-charge region* when the system reaches thermal equilibrium establishes a *built-in potential voltage* V_{BI}. Held majority carriers must therefore overcome the *energy barrier* E_B that this V_{BI} sets to diffuse across the junction:

$$E_B = q_E V_{BI},\qquad(5)$$

where q_E is the *electronic charge*, which refers to the 1.60×10^{-19} Coulombs of charge that each electron carries.

B. Energy-Band Diagram

The band gap is constant throughout the material because both P- and N-type regions are part of the same semiconductor. The probability of finding charge carriers is also uniform because net current flow is zero. Since hole and electron concentrations are high in P and N regions, respectively, E_F is closer to E_V in P material and closer to E_C in N material. So when piecing the band diagram together, E_{BG} and E_F are uniform across the device. E_F in Fig. 7 is closer to E_V in the P side than to E_C in the N side because N_A^+ is much higher than N_D. E_C in the P side is higher than in the N side because N-side electrons need additional energy to overcome the energy barrier $q_E V_{BI}$ that impedes further diffusion.

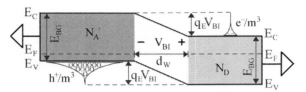

Fig. 7. Band diagram of zero-bias P^+N junction.

C. Carrier Concentrations

Hole and electron densities n_H and n_E far away from the junction and at the edge of the depletion region are uniform because they do not lose carriers to diffusion. This means that to the left of d_W in Fig. 8, where the material is P type, n_H is N_A and n_E is n_e's n_i^2/N_A. n_E is similarly N_D and n_H is n_h's n_i^2/N_D to the right of d_W, where the material is N type.

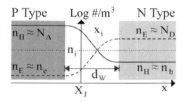

Fig. 8. Carrier densities across PN junction.

Since fewer dopant carriers reduce the propensity for thermionic carriers to recombine, P-side holes that diffuse away not only reduce n_H but also increase n_E across the junction and N-side electrons that diffuse across do the opposite. The material is intrinsic where carrier densities match (at x_i) because dopant electrons and holes neutralize. Here, thermionic emissions avail intrinsic concentrations of holes and electrons, which means n_H and n_E equal n_i.

When asymmetrically doped, the highly doped region diffuses more carriers across the junction than the lightly doped side. In the case of Fig. 8, for example, N_A^+ is much greater than N_D. So n_H near the junction X_J is greater than n_E. This is why n_H and n_E crisscross further in the N side at x_i (and not at X_J).

Interestingly, minority carrier concentration in the lightly doped side is greater than in the highly doped counterpart. This is because fewer dopants reduce the number of thermionic carriers that recombine, so more thermionic carriers survive. n_E's n_e in the P side is lower than n_H's n_h in the N side because of this: because N_A^+ is greater than N_D.

Example 2: Find n_H and n_E outside the depletion region at 27° C for a PN junction doped with $10^{17}/cm^3$ acceptor atoms and $10^{14}/cm^3$ donor atoms.

Solution:

$$n_{H(P)} \approx N_A = 10^{17} cm^{-3}$$

$$n_{E(P)} \approx n_{e(P)} = \frac{n_i^2}{N_A} = \frac{(1.45 \times 10^{10})^2}{10^{17}} = 2.10 \times 10^3 cm^{-3}$$

$$n_{H(N)} \approx n_{h(N)} = \frac{n_i^2}{N_D} = \frac{(1.45 \times 10^{10})^2}{10^{14}} = 2.10 \times 10^6 cm^{-3}$$

$$n_{E(N)} \approx N_D = 10^{14} cm^{-3}$$

Note: $n_{E(P)}$'s $n_{e(P)}$ is lower than $n_{H(N)}$'s $n_{h(N)}$ because the P side is more heavily doped than the N side, so more thermionic electrons recombine.

D. Depletion Width

More densely populated regions require less space to neutralize incoming carriers. So depletion distances from the junction are shorter when doping concentrations are higher. Total *depletion width* d_W is therefore shorter when N_A and N_D are higher:

$$d_W = d_P + d_N \propto \sqrt{\frac{1}{N_A} + \frac{1}{N_D}}. \tag{6}$$

Opposing doping concentrations across PN junctions are usually vastly different. In such cases, the highly doped region diffuses more carriers than the lightly doped side. With more incoming carriers and less neutralizing agents, depletion distance in the lightly doped side is far greater than in the highly doped region. So d_N across the P^+N junction that N_A^+ and N_D establish, for example, is usually so much longer than d_P that d_W is nearly d_N. In other words, d_W is largely the depletion distance into the lightly doped region.

Example 3: Draw the band diagram of a zero-bias PN^+ junction and approximate the relative location of the metallurgical junction X_J.

Solution: E_F in Fig. 9 is closer to E_C in the N region than to E_V in the P region because N_D^+ avails more free electrons than N_A avails holes. d_W extends more into the P region (from X_J) because the P region is lightly doped and the N region is heavily doped. So the P region requires more space to neutralize all the electrons that diffuse from the N side.

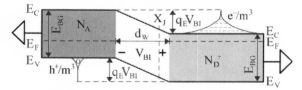

Fig. 9. Band diagram of zero-bias PN^+ junction.

E. Built-In Barrier Voltage

The energy barrier across the junction indicates that dopant electrons in the N side need E_B more energy than their equilibrium thermal energy to diffuse. The Fermi level indicates exponentially fewer electrons in the conduction band carry higher energy. So when combined, $n_{e(P)}$ is lower than N_D by the exponential amount that E_B and V_{BI} overwhelm the thermal energy E_T and *thermal voltage* V_t that *junction temperature* T_J establishes:

$$n_{E(P)} \approx n_{e(P)} = \frac{n_i^2}{N_A} = N_D \exp\left(\frac{-E_B}{E_T}\right)$$

$$= N_D \exp\left(\frac{-q_E V_{BI}}{K_B T_J}\right) = N_D \exp\left(\frac{-V_{BI}}{V_t}\right),$$

(7)

where K_B is *Boltzmann's constant* 1.38×10^{-23} J/K, T_J is in Kelvin K, E_T is $K_B T_J$, and V_t is $K_B T_J / q_E$. V_{BI} is therefore a V_t and a logarithmic translation of how much N_A and N_D dwarf n_i^2:

$$V_{BI} = V_t \ln\left(\frac{N_A N_D}{n_i^2}\right).$$

(8)

Example 4: Find V_{BI} at 27° C for a PN junction doped with 10^{17}/cm³ acceptor atoms and 10^{14}/cm³ donor atoms.

Solution:

$$V_{BI} = V_t \ln\left(\frac{N_A N_D}{n_i^2}\right) = \left(\frac{K_B T_J}{q_E}\right) \ln\left(\frac{N_A N_D}{n_i^2}\right)$$

$$= \left[\frac{(1.38 \times 10^{-23})(300)}{(1.60 \times 10^{-19})}\right] \ln\left[\frac{(10^{17})(10^{14})}{(1.45 \times 10^{10})^2}\right]$$

$$= (25.9 \text{ mV}) \ln (0.70 \times 10^{11}) = 650 \text{ mV}$$

2.2. Reverse Bias

A. Electrostatics

<u>Dopant Carriers</u>: Applying a negative voltage across the PN junction reinforces the built-in electric field. This effectively raises the energy

barrier that majority carriers must overcome to diffuse, so dopant carriers do not diffuse. This *reverse voltage* v_R in Fig. 10 increases the barrier to $q_E(V_{BI} + v_R)$.

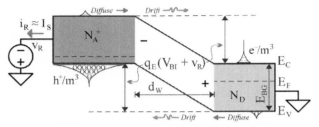

Fig. 10. Band diagram of reverse-bias PN junction.

Thermionic Carriers: Since the higher barrier deactivates dopant carriers, thermionic emissions avail more minority carriers than dopants in the other side of the junction avail majority carriers. In other words, P-side electrons outnumber N-side electrons and N-side holes outnumber P-side holes. Thermionic carriers therefore diffuse, and with the reverse field that V_{BI} and v_R set, drift across the depletion region. This flow of thermionic carriers establishes a *reverse current* i_R.

B. I–V Translation

i_R peaks and saturates quickly with increasing v_R because thermal energy in semiconductors liberates few electrons. This is why *reverse saturation current* I_S is normally low and why a v_R of only three V_t's raises i_R to 95% of the I_S that is possible:

$$i_R = I_S \left[1 - \exp\left(\frac{-v_R}{n_I V_t} \right) \right]. \tag{9}$$

n_I is the *non-ideality factor* used to compensate for second-order effects. Structural imperfections in the depletion region, for example, can trap mobile electrons. Since this reduces the effect of v_R, n_I is usually higher (up to two) than in the ideal case, for which n_I is one.

Several components dictate I_S's charge rate dq_S/dt_S. Electronic charge q_E is the most basic of these. Charge-carrier concentrations are next. I_S is proportional to cross-sectional *junction area* A_J, for example, because larger areas supply more carriers:

$$I_S \equiv \frac{dq_S}{dt_S} = q_E A_J \left(n_{e(P)} \sqrt{\frac{D_N}{\tau_N}} + n_{h(N)} \sqrt{\frac{D_P}{\tau_P}} \right)$$
$$= q_E A_J \left[n_{e(P)} \left(\frac{n_i^2}{N_A} \right) \left(\frac{D_N}{L_N} \right) + \left(\frac{n_i^2}{N_D} \right) \left(\frac{D_P}{L_P} \right) \right].$$

(10)

I_S is similarly proportional to thermionic carrier concentration: minority electron density $n_{e(P)}$ in the P region and minority hole density $n_{h(N)}$ in the N region and the junction temperature that produces them. These minority carrier concentrations ultimately hinge on doping concentrations N_A and N_D.

The number of carriers that cross also depends on *diffusivity*, which is the ability of charge carriers to diffuse. Electron and hole *diffusion coefficients* D_N and D_P quantify this effect. Charge rate also depends on the time that traversing carriers require to recombine. I_S is therefore higher when N- and P-type *average carrier lifetimes* τ_N and τ_P are shorter. Another way to describe the temporal effect of τ_N and τ_P is spatially with *average diffusion lengths* L_N and L_P because L_N and L_P are, respectively, square-root translations of $D_N \tau_N$ and $D_N \tau_N$.

C. Depletion Width

Note that v_R applies a negative voltage to the P region and a positive voltage to the N region. So v_R attracts dopant holes and electrons away from the junction. v_R therefore widens the depletion region and the depletion distances that quantify the separation: d_P, d_N, and d_W.

2.3. Breakdown

A. Impact Ionization

When the reverse voltage is very high, v_R accelerates thermionic electrons to such an extent and with such kinetic force that they collide and liberate bound electrons. So one energized electron in Fig. 11 drifts, collides, and frees one electron–hole pair that avails another electron and a hole. v_R energizes the two liberated electrons to the same degree, so they generate two other electrons and two other holes. This multiplicative action continues as long as v_R is above the *breakdown voltage* V_{BD} that induces it.

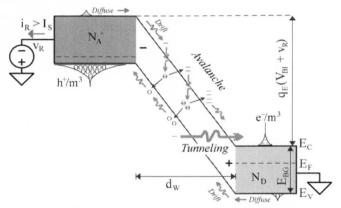

Fig. 11. Band diagram of PN junction in reverse breakdown.

This process of colliding and releasing built-up energy to liberate electrons is *impact ionization*. Since reverse current builds and grows avalanche-fashion, this phenomenon is known as *avalanche breakdown*. The breakdown voltage for this mechanism is higher when doping concentrations are lower because, with the wider depletion regions that result, field intensity is lower. Typical V_{BD}'s are greater than 5 V.

B. Tunneling

When v_R is high and depletion width is very narrow, the field is so intense that valence electrons in the P region in Fig. 11 break away and

Reverse

tunnel through the depletion region. This is *zener tunneling*. Reverse current climbs above I_S this way as long as v_R is higher than the V_{BD} that induces i_R.

For the depletion width to be so narrow, doping concentrations must be very high. This is why *zener breakdown* normally happens across highly doped junctions. Typical V_{BD}'s are less than 7 V.

C. Convention

Irrespective of which mechanism dominates, engineers normally call diodes optimized to operate in breakdown *zener diodes*. Interestingly, many of these diodes "break" around 6–7 V. At this level, i_R is the result of both avalanche and tunneling effects. Note, by the way, that neither breakdown mechanism is destructive.

2.4. Forward Bias

A. Electrostatics

Applying a positive *diode voltage* v_D across the PN junction opposes the built-in electric field. This effectively reduces the energy barrier that dopant carriers must overcome to diffuse to the $q_E(V_{BI} - v_D)$ that Fig. 12 shows. As v_D reduces E_B, an exponentially increasing number of dopant carriers become available.

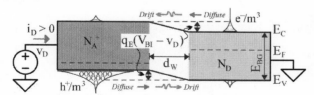

Fig. 12. Band diagram of forward-bias PN junction.

Since doping concentrations are so much higher than thermionic concentrations, dopant carriers quickly outnumber minority (thermionic) carriers in the opposing regions. The resulting concentration difference actuates diffusion of dopant carriers into the junction. The electric field present sweeps these dopant carriers across the depletion region to establish a *forward diode current* i_F or i_D.

Note that diffusing electrons penetrate the P region and diffusing holes enter the N region to establish i_D. So electrons become minority carriers in the P region and holes become minority carriers in the N region. In other words, i_D is the result of minority-carrier conduction.

B. I–V Translation

i_D is zero with zero bias. i_D rises when incoming carriers outnumber thermionic carriers. Since I_S is thermionic current and raising v_D avails an exponential number of diffusing carriers, i_D is a scalar translation of I_S that is zero when v_D is zero and increases exponentially with v_D:

$$i_D = I_S\left[\exp\left(\frac{v_D}{n_I V_t}\right) - 1\right] \propto A_J. \qquad (11)$$

Doping concentration is so high that three V_t's can establish an i_D that is 20 times greater than the thermionic current that limits i_R to I_S.

When v_D overcomes the built-in potential, the barrier fades and the depletion region shrinks. Since little impedes diffusion, the diode becomes practically a short. So i_D skyrockets past the "knee" that V_{BI} sets.

Non-ideality: The non-ideality factor when i_D is low is similar to i_R's (greater than one) because imperfections trap a substantial fraction of diffusing electrons. All traps eventually fill, however, so higher current reduces the fraction of electrons lost in traps. Since diffusing electrons increase exponentially with v_D, n_I falls as v_D climbs and approaches one when v_D is roughly $0.5V_{BI}$ to V_{BI}, when diffusion overwhelms second-order effects.

Near and above V_{BI}, incoming carriers can outnumber dopant carriers, so fewer carriers recombine and n_I is again greater than one.

This *high-level injection* occurs first in the region with the least doping concentration. With these higher current levels, the bulk regions and their contacts drop an ohmic voltage that compresses v_D. When this happens, i_D reduces to a linear translation of the external voltage applied.

C. Depletion Width

Note that v_D applies a positive voltage to the P region and a negative voltage to the N region. So v_D repels dopant holes and electrons into the junction. v_D therefore narrows the depletion region and the depletion distances that quantify the separation: d_P, d_N, and d_W.

2.5. Model

A. Symbol

The PN junction conducts substantial current when forward-biased and hardly any current when reverse-biased. So the symbol that represents it in Fig. 13 is an arrow that points in the direction of forward current i_D. To highlight that current does not reverse (by much), a blocking line crosses the tip of the arrow. i_D enters the *anode* terminal of the diode and exits from the *cathode* terminal.

Fig. 13. PN diode symbol.

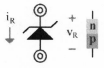

Fig. 14. Zener diode symbol.

Diodes optimized to operate in breakdown receive the same basic symbol, because the overall behavior is the same. But since breakdown conducts substantial reverse current, the blocking line in Fig. 14 flares out. These "wings" essentially indicate that the blocking mechanism is conditional.

B. I–V Translation

Notice that the equation for i_D matches i_R when i_D and v_D are negative. So the expression describes both forward and reverse conditions, but not breakdown. So since i_D is high and negative in breakdown, i_D in Fig. 15 skyrockets down near $-V_{BD}$, saturates to $-I_S$ in reverse bias, increases exponentially with v_D in forward bias, and skyrockets up near V_{BI}.

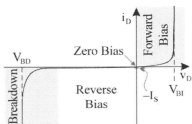

Fig. 15. Diode's Current–voltage translation.

The diode is practically a short in breakdown and at V_{BI} and an open circuit otherwise. This is why engineers often use them as on–off switches. When used this way, the diode switch closes and drops close to V_{BI} with forward current and opens whenever current reverses.

C. Dynamic Response

<u>Small Variations</u>: Shifting the operating point of a diode requires charge. Raising the barrier voltage, for example, repels recombined carriers back to their home regions. The number of carriers that diffuse across the junction also decreases. Moving these carriers changes the charge concentration across the junction.

This requires time because i_D carries a finite amount of charge per second. So the current and charge needed Δi_D and Δq_D to vary the voltage Δv_D dictate the response time Δt_R of the diode. *Junction capacitance* C_J, which is the charge held across the junction with one volt, relates these parameters:

$$C_J = \frac{\Delta q_D}{\Delta v_D} = \frac{\Delta i_D \Delta t_R}{\Delta v_D} = C_{DEP} + C_{DIF}. \tag{12}$$

Depletion capacitance C_{DEP} is the component that the depletion region holds. *Diffusion capacitance* C_{DIF} is the diffusion charge held in transit.

The depletion region is void of charge carriers and non-conducting, like an insulator, with the P and N regions as ohmic contacts. This parallel-plate structure is what establishes C_{DEP}. C_{DEP} therefore increases with junction area and field intensity, and as a result, with A_J/d_W:

$$C_{DEP} = \frac{C_{J0}}{\sqrt{\dfrac{V_{BI} - v_D}{V_{BI}}}} = \frac{A_J C_{J0}"}{\sqrt{1 - \dfrac{v_D}{V_{BI}}}} \propto \frac{A_J}{d_W}. \tag{13}$$

Since a positive diode voltage narrows d_W, C_{DEP} also increases with v_D.

Doping concentration determines how many charge carriers are available and the barrier voltage $V_{BI} - v_D$ how many of those vanish in the depletion region. So C_{DEP} not only rises with v_D but also with N_A and N_D. Since diffusion current disappears with zero bias, measuring C_J when v_D is zero to determine the *zero-bias junction capacitance* C_{J0} and using C_{J0} to extrapolate the effect of v_D in $V_{BI} - v_D$ on C_J is a practical way of quantifying C_{DEP}. Normalizing C_{J0} to area with $C_{J0}"$ is even better because A_J is a design variable.

Forward-biased carriers diffuse across the junction to become minority carriers. If the doped regions are short, these carriers in Fig. 16 can reach the metallic contacts without recombining. Irrespective, diffusing carriers require *forward transit time* τ_F to feed i_D. The voltage that sets this i_D dictates the number of in-transit charge q_{DIF} "held" by this mechanism.

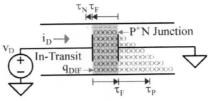

Fig. 16. In-transit diffusion charge when forward-biased.

C_{DIF} is the charge needed Δq_{DIF} for the voltage to vary Δv_D:

$$C_{DIF} = \frac{\Delta q_{DIF}}{\Delta v_D} = \frac{\Delta i_D \tau_F}{\Delta v_D}\Bigg|_{\text{Small Variations}} \approx \left(\frac{\partial i_D}{\partial v_D}\right)\tau_F \approx \left(\frac{I_D}{V_t}\right)\tau_F \equiv g_m \tau_F. \quad (14)$$

Since Δi_D after τ_F avails Δq_{DIF}, C_{DIF} is sensitive to i_D and the v_D that sets i_D. C_{DIF} is nonlinear because i_D is an exponential translation of v_D. For small variations, however, $\Delta i_D/\Delta v_D$ is roughly the linear translation that i_D's partial derivative $\partial i_D/\partial v_D$ or I_D/V_t sets. This derivative is the *small-signal transconductance* g_m of the diode, where I_D is the static (non-varying) component of i_D.

Since all carriers ultimately recombine in long diodes, τ_F is their average lifetime. τ_F in asymmetrically doped junctions is the average lifetime of the dominant carrier. So τ_F is nearly τ_P in P$^+$N junctions and τ_N in PN$^+$ junctions. τ_F, however, is a fraction of that in short diodes, in which case C_{DIF} is lower. In other words, short diodes are fast.

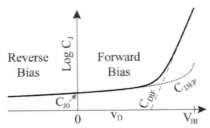

Fig. 17. Junction capacitance.

C_{DIF} climbs exponentially with v_D because diffused carriers in i_D increase exponentially with v_D. So C_{DIF} in Fig. 17 overwhelms C_{DEP} 200–

400 mV before v_D reaches V_{BI}. In other words, C_J is practically C_{DEP} in reverse bias and light forward bias and C_{DIF} when v_D equal to or greater than 300–500 mV.

<u>Large Transitions</u>: When used as switches, diodes transition between the on and off states that forward- and reverse-bias conditions establish. *Reverse-recovery time* t_{RR} refers to the time needed to reverse-bias the junction from a forward-bias state. t_{RR} is therefore the time needed to pull in-transit diffusion carriers q_{DIF} back to their home regions and to recombine carriers q_{DEP} in the depletion region. But of these, q_{DIF} is usually much greater than q_{DEP}.

t_{RR} is largely the time that reverse current i_R requires to collect the *reverse-recovery charge* q_{RR} that q_{DIF} sets:

$$t_{RR} = \frac{q_{RR}}{i_R} = \frac{q_{DIF} + q_{DEP}}{i_R} \approx \frac{q_{DIF}}{i_R} \approx \left(\frac{i_F}{i_R}\right)\tau_F, \qquad (15)$$

where q_{DIF} is the charge that forward-bias current i_F produces with τ_F and i_R can vary with time. So t_{RR} ultimately hinges on i_F and the i_R that circuit conditions avail. When unchecked, i_R can peak to a level $i_{R(PK)}$ in Fig. 18 that is comparable to and possibly higher than $-i_F$. When limited to a lower level $i_{R(LIM)}$, $i_{R(LIM)}$ extends t_{RR}. This is unfortunate either way because the diode should be off, not conducting this much reverse current.

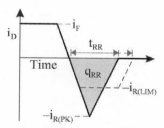

Fig. 18. Reverse-recovery current.

Forward-recovery time t_{FR} refers to the time needed to forward-bias a reverse-biased junction. t_{FR} is therefore the time needed to supply in-transit diffusion and depletion carriers. Since q_{DIF} is much greater than q_{DEP}, *forward-recovery charge* q_{FR} is nearly q_{DIF}.

This transition is more benign than reverse recovery in two ways. First, the forward current i_F needed to supply q_{DIF} flows from anode to cathode, as a diode should. Second, the circuit avails the i_F that sets q_{DIF} in the first place:

$$t_{FR} = \frac{q_{FR}}{i_F} = \frac{q_{DIF} + q_{DEP}}{i_F} \approx \frac{q_{DIF}}{i_F} \approx \frac{i_F \tau_F}{i_F} = \tau_F. \tag{16}$$

So t_{FR} is nearly the transit time of the diode.

3. Metal–Semiconductor (Schottky) Diodes

Electrons in metal are so weakly bound that they are practically free and available for conduction. Still, the probability of finding electrons above the Fermi level of metal E_{FM} decreases exponentially with electron energy E_E. The Fermi level E_{FS} of a semiconductor is greater than E_{FM} when the semiconductor has more high-energy electrons, like the N semiconductor in Fig. 19.

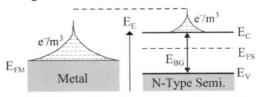

Fig. 19. Band diagram of separate metal and N-type semiconductor solids.

3.1. Zero Bias

Since carrier concentration at higher energy levels is higher in the semiconductor, electrons diffuse into the metal when connected together. Diffusing electrons deplete and ionize the semiconductor region near the junction X_J. Electrons continue to diffuse until the growing electric field is high enough to repel further action.

Since current cannot flow under zero-bias conditions, the probability of finding charge carriers E_{FM} and E_{FS} in Fig. 20 is uniform across the junction. Higher electron density in the semiconductor $n_{E(S)}$ induces more diffusion. As a result, the semiconductor ionizes more and the built-in voltage V_{BI} that the barrier establishes is higher. The depletion width is narrower when $n_{E(S)}$ is higher because depleting a region that is more dense with electrons is more difficult.

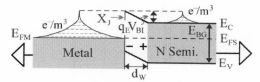

Fig. 20. Band diagram of zero-bias N-type metal–semiconductor junction.

3.2. Reverse Bias

Applying a positive voltage v_R to the semiconductor raises the barrier voltage $V_{BI} + v_R$ that electrons in the semiconductor need to overcome and diffuse into the metal. So no electrons diffuse. Still, v_R pulls thermionic electrons in the metal into the semiconductor. But since thermionic electron density is low, reverse current quickly saturates to the reverse-saturation current of the junction.

3.3. Forward Bias

Applying a positive voltage v_D to the metal does the opposite: reduces the barrier voltage $V_{BI} - v_D$ that electrons in the semiconductor need to overcome and diffuse into the metal. So electrons diffuse and establish current flow. Since the number of high-energy electrons climbs exponentially with a lower barrier voltage, i_D increases exponentially with v_D.

3.4. Model

A. Symbol

The metal–semiconductor junction conducts substantial i_D when forward-biased and hardly any i_R when reversed. So like the PN diode, the symbol that represents the metal–semiconductor junction in Fig. 21 is an arrow that points in the direction of i_D. To highlight that current does not reverse (by much), a blocking line crosses the tip of the arrow. But to distinguish it from the PN diode, the ends of the blocking line square back. Engineers often call this structure a *Schottky* or *Schottky-Barrier diode* after the physicist recognized for this junction. *Hot-electron* and *hot-carrier diodes* are also common appellations because diffusing electrons carry more energy than electrons in the metal.

Fig. 21. Metal–semiconductor (Schottky) diode symbol.

B. I–V Translation

i_D rises so much when v_D reaches V_{BI} that the junction practically shorts, like Fig. 15 shows. Reinforcing the barrier with a negative v_D induces so little i_R that the device opens like a switch. When the reverse voltage is high enough, however, the junction breaks down and shorts to conduct substantial i_R. So this diode behaves very much like the PN diode.

This diode, however, normally diffuses fewer carriers with zero bias than the PN junction. So the V_{BI} that these diffusing electrons establish is usually lower, about 150–300 mV. The metal also avails more thermionic electrons, so reverse current is higher when v_D reverses. Since diffusing electrons in the metal do not recombine like they would in a P region, the non-ideality factor is closer to one.

C. Dynamic Response

Carriers that diffuse away and vanish from the depletion region ionize the junction. Since the barrier voltage $V_{BI} - v_D$ keeps more carriers from diffusing, reducing v_D reduces the number of carriers that vanish and the charge they establish. Depletion capacitance C_{DEP} therefore holds less charge when v_D falls and reverses, like in the PN diode. C_{DEP} also scales with doping concentration because more carriers diffuse when carrier density is higher.

Unlike PN diodes, however, forward-bias electrons do not traverse a P region before reaching a metallic contact. And no holes diffuse into the semiconductor. So carriers do not require forward transit time τ_F to feed i_D. This means that the process of supplying and retrieving electrons is nearly instantaneous. And, in-transit charge and diffusion capacitance are nil. With no q_{DIF} to recover, reverse-recovery time is short.

D. Diode Distinctions

Metal–semiconductor diodes are faster than PN diodes. Plus, they drop lower voltages, so they burn less power. The drawback is, they also leak more reverse current. All this is because diffused electrons are majority carriers in metal and minority carriers in a P region. So forward i_D is the result of majority-carrier conduction in Schottkys and minority-carrier conduction in PN diodes. In a way, Schottkys behave like P⁻N junctions with very short P regions because diffusing electrons alone establish V_{BI} and feed i_D and τ_F in the P region is very short.

3.5. Structural Variations

A. P Type

The P-type Schottky diode is the complement of the N type. When E_{FS} in Fig. 22 is below E_{FM} and the two materials connect, semiconductor holes need less energy than metal electrons to diffuse. So metal electrons populate and pull semiconductor holes into the metal, ionizing the semiconductor until the field that builds is high enough to keep other holes from diffusing.

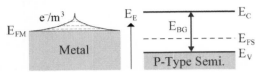

Fig. 22. Band diagram of separate metal and P-type semiconductor solids.

V_{BI} in Fig. 23 is the barrier that keeps other holes from diffusing. The number of holes that diffuse and vanish from the depletion region is usually so low that V_{BI} is very low. So the reverse leakage current that results is correspondingly high. This is why N-type Schottkys are more prevalent in practice, because they are easier to optimize (for lower leakage).

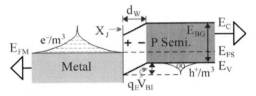

Fig. 23. Band diagram of zero-bias P-type metal–semiconductor junction.

B. Contacts

The depletion region is so narrow when the N semiconductor is highly doped that electrons can tunnel easily across. As a result, both positive and negative voltages across the junction induce substantial forward and reverse currents. This way, the junction forms a *Schottky contact*.

When high-energy electron density in the metal E_{FM} is higher than in the N semiconductor E_{FS}, metal electrons diffuse into the semiconductor with zero bias. When this happens, electrons *accumulate* near the junction. So electrons can flow easily in both directions. This way, the junction is an *ohmic contact*.

4. Bipolar-Junction Transistors

The *bipolar-junction transistor* (BJT) is basically two diodes head-to-head or back-to-back where the sandwiched "head" or sandwiched "back" is narrow. The transistor is "bipolar" because the structure is symmetrical, so the transistor can steer current in both directions. Engineers call the middle region the *base* because, when first built, it also served as the mechanical base of the structure.

4.1. NPN

In an NPN, head-to-head diodes sandwich a thin P base like Fig. 24 shows. The short distance between the junctions is the *metallurgical base width* W_B. The *effective base width* w_B is shorter because base holes near the junctions diffuse away into the N regions. So the depletion regions effectively squeeze the base.

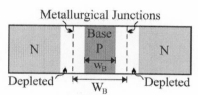

Fig. 24. NPN BJT structure.

A. Active

<u>Bias</u>: With a short base, the BJT *activates* when one diode forward-biases and the other diode reverses. In Fig. 25, the *base voltage* v_B forward-biases the junction on the left with respect to v_E and reverse-biases the junction on the right with respect to v_C. So v_C is greater than v_B, which is in turn greater than v_E. As a result, the field barrier and depletion region on the left decrease and the counterparts on the right increase.

<u>Electrostatics</u>: Electrons and holes therefore diffuse across the junction on the left. w_B is so short that almost all diffused electrons reach the

opposite end of the base without recombining. Since v_C is greater than v_B, v_C pulls these base electrons across the depletion region into the N region on the right.

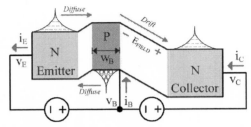

Fig. 25. Band diagram of NPN when activated.

This way, the N region on the left "emits" the electrons that the N region on the right "collects". And as this happens, the P base injects holes into the N *emitter*. So the emitter receives the electron and hole currents i_{E-} and i_{E+} that the *collector* and base supply with *collector* and *base currents* i_C and i_B:

$$i_E = i_{E-} + i_{E+} = i_C + i_B. \tag{17}$$

<u>Determinants</u>: Of *emitter current* i_E, i_C loses i_{E+} to i_B. This i_{E+} is largely the fraction of i_{E-} that the *base* and *emitter doping concentrations* N_B and N_E and effective diffusion distance of electrons in the base w_B (because w_B is shorter than the average diffusion length of electrons across a long base L_{B-}) and average diffusion length L_{E+} of holes in the emitter set:

$$\frac{i_{E+}}{i_{E-}} = \left(\frac{N_B D_{E+}}{L_{E+}}\right)\left(\frac{w_B}{N_E D_{B-}}\right) \propto \left(\frac{N_B}{N_E}\right)\left(\frac{w_B}{L_{E+}}\right), \tag{18}$$

along with diffusivity of holes in the emitter D_{E+} and diffusivity of electrons in the base D_{B-}. So *emitter injection efficiency* γ_E, which is the fraction of i_E that i_{E-} avails, is mostly an $N_E L_{E+}$ fraction of $N_B w_B$ and $N_E L_{E+}$:

$$\gamma_E \equiv \frac{i_{E-}}{i_E} = \frac{i_{E-}}{i_{E+} + i_{E-}} \propto \frac{N_E L_{E+}}{N_B w_B + N_E L_{E+}}. \tag{19}$$

But not all electrons in i_{E-} reach the collector. Some of them recombine with holes in the base. And more recombine when w_B is a larger fraction of L_{B-}. The *recombination current* i_{RC} that results is a quadratic w_B/L_{B-} fraction of i_{E-}:

$$i_{RC} \approx 0.5 i_{E-} \left(\frac{w_B}{L_{B-}} \right)^2. \tag{20}$$

So the *base-transport factor* α_T, which is the fraction of i_{E-} that feeds i_C, is

$$\alpha_T \equiv \frac{i_C}{i_{E-}} = \frac{i_{E-} - i_{RC}}{i_{E-}} \approx 1 - 0.5 \left(\frac{w_B}{L_{B-}} \right)^2. \tag{21}$$

Notice that α_T is nearly 90% when w_B is 45% of L_{B-}.

i_{RC} and α_T also depend on *collector voltage* v_C because v_C sets the width of the depletion region that squeezes w_B. So a higher v_C expands the depletion region, which in turn shrinks w_B, reduces i_{RC}, and raises α_T. w_B is therefore shorter than the unbiased base width w_{B0} (when *emitter voltage* v_E, v_B, and v_C are zero). The ultimate effect of this *base-width modulation* is a linear variation in i_C. Since depleting the region near the junction is easier when lightly doped, this effect is more severe when *collector doping concentration* N_C and N_B are lower.

Current Translations: γ_E and α_T set the BJT's overall *baseline transport factor* α_0:

$$\alpha_0 \equiv \frac{i_C}{i_E} = \gamma_E \alpha_{T0} \propto \left(\frac{N_E L_{E+}}{N_B w_B + N_E L_{E+}} \right) \left[1 - 0.5 \left(\frac{w_{B0}}{L_{B-}} \right)^2 \right]. \tag{22}$$

This α_0 is nearly 100% when N_E is much greater than N_B and w_{B0} is much shorter than L_{E+} and L_{B-}. Note that α_0 is the emitter-to-collector current-gain translation.

α_0 is close to 100% when i_C loses little to i_B. This is another way of saying that the base-to-collector *baseline current gain* β_0 is high. Since N_B and N_E dictate how i_E splits into i_{E-} and i_{E+} and w_B/L_{B-} determines how much of i_{E-} is lost to i_{RC}, w_{B0}/L_{B-} limits the gain that N_E/N_B sets:

$$\beta_0 \equiv \frac{i_C}{i_B} = \frac{i_{E-} - i_{RC}}{i_{E+} + i_{RC}} = \frac{i_{E-} - 0.5 i_{E-}\left(w_{B0}/L_{B-}\right)^2}{i_{E+} + 0.5 i_{E-}\left(w_{B0}/L_{B-}\right)^2}$$

$$= \left(\frac{N_E}{N_B}\right)\left\{\frac{1 - 0.5\left(w_{B0}/L_{B-}\right)^2}{\left[\left(D_{E+}w_B\right)/\left(L_{E+}D_{B-}\right)\right] + 0.5\left(N_E/N_B\right)\left(w_{B0}/L_{B-}\right)^2}\right\}. \tag{23}$$

α_0 and β_0 relate because i_E in α_0 feeds i_B in β_0 and i_C in α_0 and β_0:

$$\alpha_0 \equiv \frac{i_C}{i_E} = \frac{i_C}{i_B + i_C} = \frac{1}{i_B/i_C + 1} = \frac{\beta_0}{1 + \beta_0}. \tag{24}$$

α_0 and β_0 are ultimately measures of quality in a BJT that improve when w_B is shorter (because W_B is shorter, N_B is lower, and v_C is higher).

Collector Current: The i_{E-} that i_C collects is the forward-bias diode current that v_B and v_E establish with *base–emitter voltage* v_{BE} or $v_B - v_E$:

$$i_C \approx i_{E-}\left(1 + \frac{v_{CE}}{V_A}\right) \approx I_S\left[\exp\left(\frac{v_{BE}}{n_I V_t}\right) - 1\right]\left(1 + \frac{v_{CE}}{V_A}\right) \propto A_{JBE}. \tag{25}$$

i_C is therefore proportional to I_S, and in consequence, to the cross-sectional area of the base–emitter junction A_{JBE}. But since raising v_C narrows w_B, which in turn increases the fraction of i_{E-} that feeds i_C, i_C climbs with *collector–emitter voltage* v_{CE}. V_A is the process-dependent constant that models this linear effect that base-width modulation produces. Engineers call this parameter *Early Voltage* after the scientist that first observed and modeled this behavior.

B. Saturation

The BJT *saturates* when v_B forward-biases both junctions. As a result, both barrier voltages and depletion regions decrease and charge carriers in Fig. 26 diffuse across both junctions. Electron densities at the edges of the base match when the junctions forward-bias equally. So instead of diffusing across the base, all emitter and collector electrons recombine with base holes. As a result, v_B feeds current to both N regions.

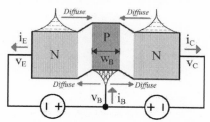

Fig. 26. Band diagram of symmetrically biased NPN in saturation.

When one junction is less forward-biased than the other, fewer electrons diffuse across this junction. So the difference diffuses across the base to the least forward-biased end (to the right in Fig. 27). A fraction of these electrons recombines with the few base holes that diffuse in the same direction. So when v_{BE} exceeds the *base–collector voltage* v_{BC}, the electrons that survive establish an i_C that flows into v_C, and together with i_B, flow out of v_E as i_E.

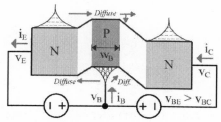

Fig. 27. Band diagram of asymmetrically biased NPN in saturation.

<u>Current Translations</u>: Collector electrons that diffuse into the base diminish electron diffusion across the base, so i_C collects fewer electrons. Decreasing N_C reduces this effect. i_C also loses electrons to forward-

biased base holes, so i_{RC} falls when N_B is lower. i_C loses so much to these effects when deeply saturated that i_C can reverse direction. But even if only lightly saturated, α_0 and β_0 are still lower in saturation than in the active region.

C. Optimal BJT

The emitter injects more diffusion current that the collector receives when N_E is higher than N_B. Plus, the collector loses less diffusion current to the base when N_C is lower than N_B. So α_0 and β_0 are greater when $N_E > N_B > N_C$. But reducing N_B and N_C also increases base-width modulation, which sensitizes i_C to v_C with a lower V_A. The optimal BJT therefore reduces N_B and N_C only to the extent that an acceptably high V_A allows. Typical values for the α_0, β_0, and V_A of an optimized N^+PN^- BJT are 98%–99% A/A, 50–70 A/A, and 50–100 V.

D. Symbol

Engineers design NPNs so collectors receive most of the electrons that emitters inject. This way, when active, i_B is very low. The orthogonal wall-like line into which i_B in Fig. 28 flows represents the base of the NPN for this reason, because a short base effectively blocks i_B.

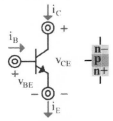

Fig. 28. NPN BJT symbol.

In this mode, only the base–emitter junction forward-biases. So v_{BE} in an NPN is positive and i_B flows into the base and out of the emitter. To illustrate this diode-like behavior, an arrow between the base and emitter terminals points in the direction of i_B: toward the emitter.

E. Modes

<u>Direction</u>: Given their symmetry, BJTs are bi-directional. As long as one diode forward-biases and the other reverses, the BJT is active. When saturated, the BJT favors the junction with the highest forward-bias current.

<u>Orientation</u>: When asymmetrically doped, the end with the highest doping concentration can inject more carriers into the base that the other side can collect. This highly doped terminal is therefore more optimal as an emitter than a collector. So by convention, the "emitter" usually refers to the highest doped end. So in an N^+PN^-, the N^+ terminal is normally the emitter and the N^- terminal is the collector.

<u>Forward Active</u>: The BJT is forward active when the base–emitter junction forward-biases and the base–collector junction reverses. So in the NPN, v_{BE} is positive and v_{BC} is negative. In other words, v_{CE} is higher than v_{BE} and i_C in Fig. 29 is exponential with v_{BE} and linear with v_{CE} when the BJT is forward active.

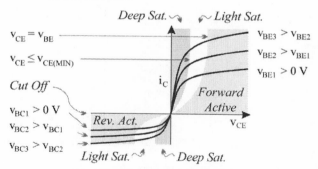

Fig. 29. N^+PN^- collector current.

<u>Forward Saturation</u>: The BJT saturates when both junctions forward-bias. So the forward-active NPN saturates when v_{CE} falls below v_{BE}. Since the transition happens when v_{CE} matches v_{BE}, i_C along the active–saturation boundary climbs exponentially with the v_{BE} that v_{CE}'s saturation point establishes.

When the base–collector junction forward-biases by less than 200 mV or so, base–emitter diffusion is so much greater that the effects of base–collector diffusion are negligible. So i_C remains exponential with v_{BE} and linear with v_{CE}. In other words, *light saturation* is largely an extension of the active region.

Forward-biasing the base–collector junction by more than 200 mV or so diffuses so many electrons and holes that their effects are no longer negligible. Fewer electrons reach the collector, and of these, a larger fraction recombines with the base holes that diffuse in the same direction. So i_C in *deep saturation* falls appreciably with v_{CE} when v_{CE} falls below the *minimum collector–emitter voltage* $v_{CE(MIN)}$ that doping concentrations and parasitic resistances ultimately dictate. $v_{CE(MIN)}$ is oftentimes 200–400 mV.

Reverse Active: The junctions reverse roles in reverse modes. So the base–collector junction forward-biases and the base–emitter junction reverses in reverse active. In the NPN, v_{BC} is positive and v_{BE} is negative, and as a result, v_{EC} is higher than v_{BC} and i_C is exponential with v_{BC} and linear with v_{EC}.

Reverse Saturation: The reverse-active NPN saturates when the base–emitter junction forward-biases, which happens when v_{EC} falls below v_{BC}. Forward-biasing the base–emitter junction by less than 200 mV or so, however, diffuses so few carriers that their effects are negligible. So current falls appreciably with v_{EC} when v_{EC} falls below the $v_{EC(MIN)}$ that process parameters and resistances set.

Optimal Behavior: Since forward modes forward-bias the highly doped N region, forward injection efficiency is usually higher than in reverse. α_0 and β_0 are therefore greater in forward bias than in reverse under similar bias voltages. This is why the magnitude of i_C is usually higher in forward modes.

<u>Cut Off</u>: BJT currents are ultimately the result of carrier diffusion. So when both junctions reverse and diffusion stops, currents vanish. This is the *cut off* region, which happens in the NPN when v_{BE} and v_{BC} are both zero or negative. The x-axis in Fig. 29 marks this mode because i_C is zero along that line.

4.2. PNP

The PNP is the NPN's complement. In a PNP, tail-to-tail diodes sandwich the thin N base in Fig. 30. W_B is the short metallurgical distance between the junctions. w_B is the distance between the depletion regions that base electrons near the junctions leave behind when diffusing into the P regions.

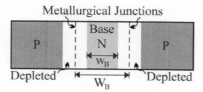

Fig. 30. PNP BJT structure.

A. Active

<u>Bias</u>: With a short base, the PNP activates when one diode forward-biases and the other diode reverses. The base voltage v_B in Fig. 31 forward-biases the junction on the right with respect to v_E and reverse-biases the junction on the left with respect to v_C. So v_E is greater than v_B, which is in turn greater than v_C. As a result, the field barrier and depletion region on the right decrease and the counterparts on the left increase.

<u>Electrostatics</u>: Electrons and holes therefore diffuse across the junction on the right. w_B is so short that almost all diffused holes reach the other

end of the base. Since v_C is lower than v_B, v_C pulls these base holes across the depletion region into the P region on the left.

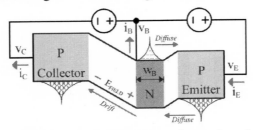

Fig. 31. Band diagram of PNP when activated.

In other words, the P region on the right "emits" the holes that the N region on the left "collects". And as this happens, the N base injects electrons into the emitter. So the emitter supplies the hole and electron currents i_{E+} and i_{E-} that the collector and base terminals receive with i_C and i_B:

$$i_E = i_{E+} + i_{E-} = i_C + i_B. \tag{26}$$

<u>Determinants</u>: Of i_E, i_C loses i_{E-} to i_B. This i_{E-} is the fraction of i_{E+} that N_B and N_E and effective diffusion distance of holes in the base w_B (because w_B is shorter than the average diffusion length of holes across a long base L_{B+}) and average diffusion length of electrons in the emitter L_{E-} set:

$$\frac{i_{E-}}{i_{E+}} = \left(\frac{N_B D_{E-}}{L_{E-}}\right)\left(\frac{w_B}{N_E D_{B+}}\right) \propto \left(\frac{N_B}{N_E}\right)\left(\frac{w_B}{L_{E-}}\right), \tag{27}$$

along with diffusivity D_{E-} of electrons in the emitter and diffusivity of holes D_{B+} in the base. So γ_E is largely an $N_E L_{E-}$ fraction of $N_E L_{E-}$ and $N_B w_B$:

$$\gamma_E \equiv \frac{i_{E+}}{i_E} = \frac{i_{E+}}{i_{E+} + i_{E-}} \propto \frac{N_E L_{E-}}{N_E L_{E-} + N_B w_B}. \tag{28}$$

Some holes in i_{E+} recombine with base electrons. And more recombine when w_B is a larger fraction of L_{B+}. i_{RC} is a quadratic w_B/L_{B+} fraction of i_{E+}:

$$i_{RC} \approx 0.5i_{E+}\left(\frac{w_B}{L_{B+}}\right)^2 . \tag{29}$$

So the fraction of i_{E-} that feeds i_C is

$$\alpha_T \equiv \frac{i_C}{i_{E+}} = \frac{i_{E+}-i_{RC}}{i_{E+}} \approx 1-0.5\left(\frac{w_B}{L_{B+}}\right)^2 . \tag{30}$$

Notice that α_T is nearly 90% when w_B is 45% of L_{B+}.

i_{RC} and α_T also depend on v_C because v_C sets the width of the depletion region that squeezes w_B. So a lower v_C expands the depletion region, which in turn shrinks w_B, reduces i_{RC}, and raises α_T. w_B is therefore shorter than the unbiased base width w_{B0} (when v_E, v_B, and v_C match). This base-width modulation produces a linear variation in i_C that is more severe when the region near the junction is easier to deplete, which happens when N_B and N_C are lower.

<u>Current Translations</u>: γ_E and α_T set the overall baseline α_0 of the PNP:

$$\alpha_0 \equiv \frac{i_C}{i_E} = \gamma_E\alpha_{T0} \propto \left(\frac{N_E L_{E-}}{N_E L_{E-}+N_B w_B}\right)\left[1-0.5\left(\frac{w_{B0}}{L_{B+}}\right)^2\right]. \tag{31}$$

α_0 nears 100% when N_E is much greater than N_B and w_{B0} is much shorter than L_{E-} and L_{B+}.

α_0 is close to 100% when i_C loses little to i_B, when baseline β_0 is high. Since N_E and N_B dictate how i_E splits into i_{E+} and i_{E-} and w_B/L_{B+} determines how much i_{E+} is lost to i_R, w_{B0}/L_{B+} limits the gain that N_E/N_B sets:

$$\beta_0 \equiv \frac{i_C}{i_B} = \frac{i_{E+}-i_{RC}}{i_{E-}+i_{RC}} = \frac{i_{E+}-0.5i_{E+}\left(w_B/L_{B+}\right)^2}{i_{E-}+0.5i_{E+}\left(w_B/L_{B+}\right)^2}$$

$$= \left(\frac{N_E}{N_B}\right)\left\{\frac{1-0.5\left(w_B/L_{B+}\right)^2}{\left[\left(D_{E-}w_B\right)/\left(L_{E-}D_{B+}\right)\right]+0.5\left(N_E/N_B\right)\left(w_B/L_{B+}\right)^2}\right\}. \quad (32)$$

α_0 and β_0 are higher when w_B is shorter (because W_B is shorter, N_B is lower, and v_C is higher).

<u>Collector Current</u>: The i_{E+} that i_C receives is the forward-bias diode current that v_{EB} establishes:

$$i_C \approx i_{E+}\left(1+\frac{V_{EC}}{V_A}\right) \approx I_S\left[\exp\left(\frac{V_{EB}}{V_t}\right)-1\right]\left(1+\frac{V_{EC}}{V_A}\right) \propto A_{JBE}. \quad (33)$$

i_C is therefore proportional to I_S, and in consequence, to A_{JBE}. But since raising v_{EC} narrows the base, which in turn increases the fraction of i_{E+} that feeds i_C, i_C climbs with v_{EC}. V_A models this base-width modulation effect in i_C.

B. Saturation

The PNP saturates when both junctions forward-bias. Charge carriers therefore diffuse across both junctions. When symmetrically forward-biased, hole densities at the edges of the base in Fig. 32 match. So instead of diffusing across the base, holes recombine with base electrons. The base therefore pulls current from both P regions.

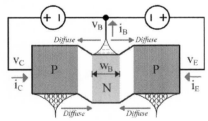

Fig. 32. Band diagram of symmetrically biased PNP in saturation.

When one junction forward-biases less than the other, fewer holes diffuse across this junction. So the difference diffuses across the base to the least forward-biased side (to the left in Fig. 33). A fraction of these holes recombines with the few base electrons that diffuse in the same direction. So when v_{EB} is greater than v_{CB}, the holes that survive establish an i_C that flows out of v_C, which is what remains of the i_E that flows into v_E after v_B pulls i_B.

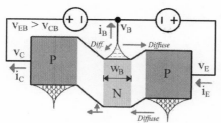

Fig. 33. Band diagram of asymmetrically biased PNP in saturation.

<u>Translations</u>: Collector holes that diffuse into the base diminish hole diffusion across the base, so i_C collects fewer holes. Decreasing N_C reduces this effect. i_C loses holes to forward-biased electrons, so i_{RC} falls with lower N_B. i_C loses so much to these effects when deeply saturated that i_C can reverse. But even if only lightly saturated, α_0 and β_0 are still lower in saturation than in the active region.

C. Symbol

Engineers design PNPs so collectors receive most of the holes that emitters inject. This way, when active, i_B is very low. The orthogonal wall-like line out of which i_B in Fig. 34 flows represents the base of the PNP for this reason, because a short base effectively limits i_B.

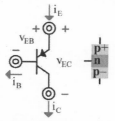

Fig. 34. PNP BJT symbol.

In this mode, only the base–emitter junction forward-biases. So in a PNP, the emitter–base voltage v_{EB} is positive and i_B flows into the emitter and out of the base. To illustrate this diode-like behavior, an arrow between the emitter and base terminals points in the direction of i_B: toward the base.

D. Modes

<u>Forward Active</u>: In forward active, the base–emitter junction forward-biases and the base–collector junction reverses. So v_{EB} is positive and v_{CB} is negative. v_{EC} is therefore higher than v_{EB} and i_C in Fig. 35 is exponential with v_{EB} and linear with v_{EC} when the PNP is forward active.

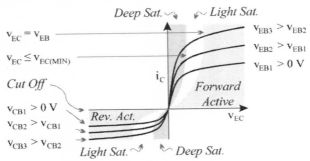

Fig. 35. P$^+$NP$^-$ collector current.

<u>Forward Saturation</u>: The forward active PNP saturates when the base–collector junction forward biases. This happens when v_{EC} falls below v_{EB}. i_C along this active–saturation boundary rises exponentially with the v_{EB} that v_{EC}'s saturation point sets. But when the base–collector junction forward-biases by less than 200 mV or so, the effects of saturation are negligible. i_C falls noticeably when this junction forward-biases by more, which happens when v_{EC} falls below $v_{EC(MIN)}$.

<u>Reverse Modes</u>: Due to its symmetry, the PNP is reversible. So reverse modes behave the same way. But when asymmetrically doped, the higher-doped end injects more charge carriers into the base than the lighter-doped side under similar conditions. So the higher-doped terminal

is, by convention, the emitter. This is why α_0, β_0, and i_C are higher in forward active and saturation (when the forward-biased base–emitter junction dominates) than in reverse active and saturation (when the forward-biased base–collector junction dominates).

Cut Off: All currents vanish when both junctions zero-bias or reverse. So i_C is zero when v_{EB} and v_{CB} are both zero or negative. The x-axis in Fig. 35 marks this mode because i_C is zero along that line.

4.3. Dynamic Response

A. Small Variations

Shifting the operating point of the diodes in the BJT requires charge q_D. And supplying or removing this q_D requires time. In the BJT, i_B and i_C supply or remove the q_D that base–emitter and base–collector junction capacitances C_{BE} and C_{BC} need. Engineers often use variables C_π and C_μ to refer to C_{BE} and C_{BC}.

Base–Emitter Capacitance: Junction capacitance C_J includes two components: the charge held across the depletion region in C_{DEP} and the charge held in transit in C_{DIF}. Since the base–emitter junction is zero- or reverse-biased in cut off, $C_{BE(DIF)}$ does not hold charge in those regions. So C_{BE} reduces to $C_{BE(DEP)}$:

$$C_{BE}\Big|_{v_{BE}<300\text{--}500\text{mV}}^{v_{BE}\le 0} \approx C_{BE(DEP)} = \frac{C_{JBE0}}{\sqrt{\dfrac{V_{BI}-V_{BE}}{V_{BI}}}} = \frac{A_{JBE}C_{JBE0}{}''}{\sqrt{1-\dfrac{V_{BE}}{V_{BI}}}}, \qquad (34)$$

where C_{JBE0} is C_{BE}'s zero-bias capacitance, A_{JBE} is the cross-sectional area of the base–emitter junction, and C_{JBE0}'' is C_{JBE0} per unit area.

Forward-biasing the base–emitter junction increases the charge that $C_{BE(DIF)}$ holds exponentially. So $C_{BE(DIF)}$ surpasses $C_{BE(DEP)}$ when v_{BE} surpasses 300–500 mV. Since v_{BE} is normally 550–700 mV in active or saturation, C_{BE} is largely the in-transit charge that $C_{BE(DIF)}$ holds and i_C supplies with v_{BE} and forward transit time τ_F across the base sets:

$$C_{BE}\Big|_{v_{BE}>300-500\,mV} \approx C_{BE(DIF)} \approx \left(\frac{\partial i_C}{\partial v_{BE}}\right)\tau_F \approx \left(\frac{I_C}{V_t}\right)\tau_F \equiv g_m\tau_F, \qquad (35)$$

Since small variations in i_C/v_{BE} are roughly the linear translation that i_C's partial derivative $\partial i_C/\partial v_{BE}$ or I_C/V_t or g_m sets, C_{BE} reduces to $C_{BE(DIF)}$'s $g_m\tau_F$.

<u>Base–Collector Capacitance</u>: Since the base–collector junction only forward-biases in saturation, $C_{BC(DIF)}$ does not hold charge when the BJT activates or cuts off. So C_{BC} in Fig. 36 reduces to $C_{BC(DEP)}$ in the active and cut-off regions when v_{CE} matches or surpasses v_{BE}:

$$C_{BC}\Big|^{v_{CE}\geq v_{BE}}_{v_{CE}\geq v_{CE(MIN)}} \approx C_{BC(DEP)}$$

$$= \frac{C_{JBC0}}{\sqrt{\dfrac{V_{BI}-V_{BC}}{V_{BI}}}} = \frac{A_{JBC}C_{JBC0}''}{\sqrt{1+\dfrac{V_{CB}}{V_{BI}}}} = \frac{A_{JBC}C_{JBC0}''}{\sqrt{1+\left(\dfrac{V_{CE}-V_{BE}}{V_{BI}}\right)}}. \qquad (36)$$

C_{JBC0} here is C_{BC}'s zero-bias capacitance, A_{JBC} is the cross-sectional area of the base–collector junction, and C_{JBC0}'' is C_{JBC0} per unit area.

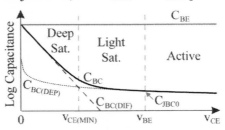

Fig. 36. BJT capacitances.

Forward-biasing the base–collector junction increases the charge that $C_{BC(DIF)}$ holds exponentially. But when lightly saturated, the effect is so minimal that $C_{BC(DEP)}$ dominates. So C_{BC} is nearly $C_{BC(DEP)}$ when v_{CE} matches or surpasses $v_{CE(MIN)}$. In deep saturation, however, $C_{BC(DIF)}$ holds more charge than $C_{BC(DEP)}$. So below $v_{CE(MIN)}$, C_{BC} is largely the in-transit charge that $C_{BC(DIF)}$ holds with v_{BC}. Note that C_{BC} is much lower than C_{BE} when lightly saturated and activated.

B. Large Transitions

BJTs activate after i_B and i_C supply the depletion and in-transit charge $q_{BE(DEP)}$ and $q_{BE(DIF)}$ that the base–emitter junction requires. To saturate, i_B and i_C must similarly supply the $q_{BC(DEP)}$ and $q_{BE(DIF)}$ that the base–collector junction requires. So when used as a switch, the BJT activates and saturates after i_B and i_C supply this charge. The BJT cuts off after i_B and i_C reverse this charge.

These large transitions require substantial i_B, and when i_B is low, substantial *recovery time* t_R because $q_{BE(DIF)}$ and $q_{BC(DIF)}$ are high. Keeping the base–collector junction from entering deep saturation reduces $q_{BC(DIF)}$ to $q_{BC(DEP)}$ levels. This way, t_R can be shorter.

A Schottky diode across the base and collector (like D_S in Fig. 37) achieves this by shunting current away from the base–collector junction. Since D_S drops less voltage than the PN junction, D_S "clamps" v_{BC} to a level that keeps the PN junction from forward-biasing too much. This way, deep saturation does not occur, so $q_{BC(DIF)}$ is always as low as or lower than $q_{BC(DEP)}$. And D_S does not cancel the resulting reduction in t_R because D_S does not hold in-transit charge.

Fig. 37. Schottky-clamped BJT.

4.4. Diode-Connected BJT

The basic difference between PN diodes and BJTs is that BJTs split the diode current into its constituent electron and hole parts. In the NPN, for example, the forward-biased base–emitter junction produces a diode current i_D that flows entirely out of the emitter. Of i_D, the collector supplies the electron component i_{D-} and the base supplies the hole component i_{D+}. Similarly, the PNP steers i_{D+} of the diffused emitter current i_D into the collector and i_{D-} into the base.

The base-to-collector connection in Fig. 38 combines and forces both parts to flow through one collector–base terminal. This connection recombines the diode current that BJTs normally split. This way, BJTs behave like diodes, inducing a diode current i_D that climbs exponentially with v_{BE} in the NPN and v_{EB} in the PNP. Engineers call this a *diode connection*. Sometimes this connection is *implicit*, like when other transistors or components connect base and collector terminals together.

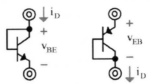

Fig. 38. Diode-connected NPN and PNP BJTs.

5. Summary

Thermal energy can liberate and avail loosely bound electrons. When exposed to an electric field, these electrons and the holes they leave behind drift in opposite directions. Conductors, semiconductors, and insulators conduct these charge carriers easily, moderately well, and poorly. Dopant atoms add electrons or holes to semiconductors so the resulting N- or P-type material avails more of one than the other. Between these, electrons flow more easily than holes.

The carrier concentration difference across a PN junction is so high that electrons and holes diffuse across, leaving behind a depleted carrier-free region. This process continues until the electric field they establish keeps other carriers from diffusing. Reinforcing the field with a reverse voltage keeps carriers from flowing. Opposing the field with a forward voltage, on the other hand, avails an exponentially increasing number of carriers for conduction. Only a negative diode current, however, can reverse these in-transit carriers.

Schottky diodes behave essentially the same way. Except, the metallic side is so full of electrons and void of holes that the metal does not deplete and holes do not diffuse into the semiconductor. Since diffusing carriers are immediately available for conduction, Schottkys are faster than PN diodes. Plus, their built-in field is usually lower, so they require a lower voltage to conduct.

BJTs are two head-to-head or back-to-back PN diodes with a short P or N base. They activate when one diode forward-biases and the other reverses. The base is so short that, biased this way, diffused minority carriers cross the entire base, where the field of the reversed diode pulls them to the collector. When the emitter's doping concentration is much higher than that of the base, the majority carriers that the base feeds is much lower than the minority carriers that the collector receives. In other words, collector current is a much greater fraction of emitter current than base current is.

Forward-biasing the second junction steers majority carriers away from the emitter. So emitter–collector and base–collector current translations fall. The reduction is lower when the collector is lightly doped and when the forward-biasing voltage across the base–collector junction is low. Reversing forward-bias charges with limited base current requires considerable time. Schottky-clamped BJTs need less time because Schottkys do not hold in-transit carriers.

Switched-inductor power supplies need transistors to energize the inductor. Although they are not always BJTs, all MOSFETs incorporate

BJTs, so understanding BJTs is essential. Plus, BJTs usually cost less money. Diodes are more basic and vital because they do not require a synchronizing signal to switch. So they can not only drain the inductor automatically but also steer current when transistors are busy transitioning states.

Made in the USA
Coppell, TX
14 November 2020

41340712R00037